Solar Energy

By Ade Asefeso MCIPS MBA

Second Edition

ISBN-13: 978-1499786958

ISBN-10: 1499786956

Publisher: AA Global Sourcing Ltd
Website: http://www.aaglobalsourcing.com

Table of Contents

4

Disclaimer

This publication is designed to provide competent and reliable information regarding the subject matter covered. However, it is sold with the understanding that the author and publisher are not engaged in rendering professional advice. The authors and publishers specifically disclaim any liability that is incurred from the use or application of contents of this book.

If you purchased this book without a cover you should be aware that this book may have been stolen property and reported as "unsold and destroyed" to the publisher. In this case neither the author nor the publisher has received any payment for this "stripped book."

Dedication

To my family and friends who seems to have been sent here to teach me something about who I am supposed to be. They have nurtured me, challenged me, and even opposed me.... But at every juncture has taught me!

This book is dedicated to my lovely boys, Thomas, Michael and Karl. Teaching them to manage their finance will give them the lives they deserve. They have taught me more about life, presence, and energy management than anything I have done in my life.

Chapter 1: What is Solar Energy?

In order to know where to find solar energy you must first know what solar energy is. Solar energy is energy from the sun. When the sun is shining solar energy is being produced as it sends the heat radiating to the earth. You can find solar energy anywhere or anything that the sun can shine on. There is a way that you can save the sunlight in order to provide heat during the cold, which is what millions of people have done throughout the years. It started thousands of years ago when people where able to use a thick lens or magnifying glass on an object that would attract the sun rays on that particular object and could get so hot it would catch fire. This gave a new prospective on how strong the heat from the sun really was.

The problem with capturing the suns heat is the fact that when it reaches the earth's surface it does not go to one particular spot. It is spread evenly over the earth where the sunlight is able to reach. When this happens you may find it difficult to heat something using only the sunlight. Although the area or object may be hot if left in the sun too long it will not reach the full heating potential of the sun's light.

In order to use the heat efficiently to heat a room after the sun goes down or when the sun is unable to shine because of clouds you will need to use a source that will attract the heat to one particular area. This source is called a solar collector. The solar collector attracts a lot of sunlight to one particular area allowing the sun to pass through the source and into the space. The objects in the space absorb and hold the heat from the sunlight and trapping it so it will not get back out with the help of the source. Glass is a great solar collector because it allows the sun to pass through it and into the space but the heat from the sun can rarely escape leaving the space under the glass to become warm or hot from the heat. The objects in the space help to hold the heat as it comes in so that the space will stay warmer longer. This allows the area to be heated using solar power.

Because glass is a natural solar collector it makes it great to put into a greenhouse or a sun room. The glass attracts the sunlight and traps the heat inside so that the temperature in the greenhouse or sun room remain warm even at night when the temperature outside may be cold.

Solar energy can be found anywhere the sun shines but in order to fill the heat you must have direct sunlight for an extended period of time. If you just use the morning sun to heat your home your house may not stay warm through the night.

Chapter 2: History of Solar Energy

Photovoltaic cells are devices that convert light into electricity. Sometimes called a solar cell, photovoltaic cells have been used since 1883. Charles Fritts invented the cell using selenium but this process was soon stopped in 1954 when Darryl Chapin used silicon instead of selenium. The photovoltaic cells use conversion through the photoelectric effect. The photoelectric effect occurs when a metallic surface is bombarded with electromagnetic radiation. Photons are the sort of electron that any electric current produces. As the light is increased, the metallic surface is converted at a slower rate at first, but more and more solar power is converted into electricity.

Every metallic surface that is used has a different threshold at which they can produce only so much energy, while the rest of the solar power is reflected after that threshold is met.

The frequency of the light and its reaction with the metallic surface cause the light to release electrons. Each photon in the light becomes a single electron of energy. The electrons that are emitted are often referred to as photoelectrons. The photoelectrons can be converted into electrical energy which can serve many uses and does not cause pollution or greenhouse gases.

At first photo voltaics cost a lot of money. The price of the energy and the equipment used to produce electricity was higher than the actual savings the user was getting. In the early days this ran at $100 per watt, these days you can usually purchase the electricity generated from the product at about $20 per watt. This kind of energy is produced in remote areas where a power line or a generator is not feasible. By placing at these cells the people that are working or living in this from an area can have a steady supply of electricity.

This type of solar cell has been used many times in public service venues. They have been used to power road crossings, road signs, street lamps, and other Department of Transportation issues that may be a long way from a source of power. It is more economically viable to put a solar cell box on a remote road crossing sign than it is to strain power chords and build power poles to get the electricity to the unit. This allows for more safety precautions being put into remote areas where the drivers or pedestrians can see the hazard.

Since the 1973 oil crisis, photovoltaics have come a long way in both technology and cost. These solar cells are even used in nautical machinery. Buoys with lights that show boats and ships the channel or the water passageway once had to be changed on a regular basis by putting in new batteries. Now these solar cells allow the buoys to remain lit from the energy of the sun. Solar power has made the roadways, sea ways, and other avenues of transportation safer and easier to navigate.

Photovoltaic modules are the prime source of power for many space systems. Solar power is the only source of energy once you leave the Earth's atmosphere. By clicking the rays of the sun on solar cells, these satellites, spacecraft, space station, and other satellite orbiting systems can remain in space longer and work with little maintenance. Without solar cells, future space explorations to other planets or other space ventures would not be possible. The capturing of the sun's rays upon a solar shield in the space station gives the astronauts electricity to run computer systems, life support systems, and everything it takes to support life while they are aboard the station.

The improved technology of solar power has increased domestic uses also. With the new trend of building greenhouses, solar panels are installed to take in the sun's energy and turn it into electricity. These solar panels have enough electricity to heat hot water tanks, run appliances, and give the family enough electricity needed to survive. Of course, supplemental electricity is provided to most of these houses in case there is a problem with the solar battery or if there is a cloudy day which would hinder the production of electricity. Solar panels have not only been used to power domestic homes, but they have also been used to power large high-rises. Entire business complexes have been powered by solar power.

The photovoltaic modules in solar cells are still being studied. They have a variety of uses that are capable of being expanded in industry, technology, space, and even in your own home. As the cells grow better they also become cheaper. Since the world's population pushes for more energy demands and the global market has raised the price of oil and other non-renewable energies to a breaking point, the average man is looking for a way to save some money. Photovoltaic cells will soon be a part of every aspect of our lives that deals with electricity. Whether it is a handheld game or an engine on an aircraft carrier, this technology will be the last remaining bastion of power once all the fossil fuels are gone or we have damaged our planet so much that fossil fuels are no longer a viable way to create energy.

Chapter 3: The uses of Solar Energy

The earth receives more energy from the sun in one hour than the power requirement of the world for the whole year. Solar energy is free, renewable, clean and sustainable. We know how to harness it and we know where to use it best. Below are the different uses of solar energy.

Grid-Tied Residential Homes

Solar energy can be used hand in hand with your current electricity provider. This is ideal to those who live in places where sunlight does not shine year round, to those who live in areas where electricity is cheap and to those who just want to use solar energy as a backup to their existing source of power. The idea is that if you have 2 sources of power, you can get uninterrupted power supply all the time.

One of the benefits of having a grid-tied solar energy system at home is that you can turn your electric meter backwards. This happens when you produce more energy than you use. The excess power that your system generates is sent out to the grid, which will be used by other households. As a result, your electric meter turns backwards and your electric provider will pay you for the amount of energy your system has produced.

Homeowners can use solar energy in producing electricity to power security lights around the house's perimeter. Since these types of lights consume as much as 5 times more power than the household's daily energy requirement, opting for solar energy system is very reasonable.

Solar power system may not be used to answer the entire power requirement of the household. Sometimes it is used to power particular equipment such as lighting, water pumping, cooking and water heating or other equipments that consume the most amount of energy.

Corporate Buildings

To save on electric bills, many companies and building owners install PV cells on their atria. On large industrial buildings, PV cells can be installed on rooftops. While the initial cash out is expensive, the amount of savings the system produces over the years will pay for itself.

Off-Grid Homes

If you have a cabin house or a farm house located far from the power grids, a solar power system can be your best source of electricity. Also, it is ideal if you need to power stand-alone sensing equipment and remote telemetry. Holiday homes that receive a substantial amount of sunlight can also benefit on this system.

Community halls, schools, clinics and other buildings that are not connected to any grid lines can use PV cells to generate power from the sun.

Signs and Street lights

Lights to brighten our street and street signs consume large amount of energy. This energy is drawn from electric providers that use conventional fossil fuels. To reduce the demand for this type of energy source, signs and street light can be installed with solar power systems that store power during the day and use it to light the streets at night. Many cities have solar panels attached to their street lights to save money and reduce fossil fuel burning.

Other Recreational Application

RVs and marine vehicles require small amount of power which are drawn from its engine. This consumes fuel and emits greenhouse gases. To reduce fuel consumption and greenhouse gases emission, owners prefer charging their batteries with solar panels.

Chapter 4: Everyday use of Solar Power

When you think about solar power, you may think of those funky looking houses that have those weird looking panels on them. That is not solar power for everyday living. Solar power has many uses beyond those of producing electricity for houses or for automobiles.

Even small electrical devices can be powered by solar energy in lieu of a battery or any outlet plug-in. Devices such as cell/mobile phones, GPS units, and laptops can be solar powered. All you need is the right connectors, the right amperage and voltage of the device and a battery to store the energy solar panels collect.

You can modify and customize your solar panels and battery device to collect any amount of energy you wish up to the threshold. After the threshold, or when the battery is full, the solar energy is just being reflected from the solar panels. The battery should be compatible with the device you want to use it for.

For smaller devices you need a smaller area with less voltage or with a different lot rating. How many times have you been out fishing or camping and your cell/mobile phone battery goes out?

Most people carry cell/mobile phone adapters for their cars so that they may recharge the battery in their car. This is fine if the engine is on or the engine has been used to charge the battery in the vehicle. But running the engine is a cause of pollutants and other toxic materials which you are releasing out into the environment just by charging up your cell/mobile phone.

By having a cell/mobile phone solar powered battery charger, you are able to take your cell phone outside and leave the solar panels out in the sun. You can even do this inside on a windowsill that is directly influenced by the sun. You don't have to worry about where to plug in your charger, you just have to worry about where the sun is and how long the solar panel needs to stay out to charge a battery. It is convenient and easy-to-use, although a little expensive.

People who work outside use all kinds of electrical equipment. From laptops to survey equipment and even for lighting, solar power can play an important role in reducing the needs for generators and batteries. If you are working at a remote site and you need power immediately, simply send out your solar panels to collect the sun's rays to produce the power that you need. Again, with this type of equipment, each solar panel needs to be equipped with the proper generator or charger that will produce electricity to either charger batteries in your device or to directly run the device.

So next time you are in your electronics store or planning a shopping trip or vacation, think about the uses of solar power and those endeavours. What device do you really want to use and how can you use solar power to have that device readily available when you need it? Solar power can operate pretty much anything you use, but you need to have the right number of solar cells and the right equipment to charge the device.

As technology advances and the demand for solar powered equipment increases, prices should go down and the technology should be accelerated. Right now only 1% of all houses in the United States and United Kingdom are powered by solar energy. If we think outside the box and go beyond the house there is an entire world of devices that can be powered by the sun. All it takes is a little imagination and a little research and someday soon we will be out of the grips of fossil fuels and non-renewable energy.

Chapter 5: Solar Energy is a Form of Green Energy Source

Solar energy is a green energy source because it is a renewable and it does not cause any harm to the environment. This is achieved by converting the sun's rays into electricity with the help of solar cells.

There are three basic approaches how we can use this form of green energy source namely passive, active and by using photovoltaic cells.

When we refer to passive solar energy, nothing is converted. What happens is the building's design helps avoid heat loss and gets the most out of day lighting.

Such a technique can also be used in homes because studies have shown that this can reduce the heating requirements by as much as 80% with minimal cost. This means you don't have to turn on the air condition or heater that often and if everyone does that, we don't consume that much electricity which we get from non-renewable resources.

The second approach which is active solar energy is the first way of converting sunlight into heat. You should know that there are certain limits to this one and all it can do is make sure you have hot water.

The third approach is the big scale version and it can power an office or an entire home. This is done with the help of solar cells that convert sunlight into electricity. The smallest ones around can be seen in calculators and watches with large ones planted over huge acres of land.

The only limitation to this green energy source is the fact that it can only generate power when the weather is good and the sun is out. Should it rain, then nothing is collected and converted. When this happens, the auxiliary system is turned on until the weather improves.

Despite that, scientists and students themselves have made solar powered cars. NASA or the National Aeronautical Space Administration has sent satellites into space that are powered by solar panels. A fully functional airport can function on its own thanks to solar power even if it is situated in the middle of the frozen desert.

So people can see the awesome power of solar energy, did you know a kilowatt of solar energy can produce 5.5 hours of electricity per day. If you have more solar cells in place, naturally you will be able to produce enough power to last several days.

Solar energy is just one form of green energy source around. Through the years, we have learned to tap other resources and these examples include wind power, geothermal energy, hydroelectricity and biogas. These are all safe and by using these more often, we don't need to depend on oil which is a non-renewable resource.

To make this happen, we have to persuade our law makers to promote the use of such resources. Although you hear speeches left and right about their concern for the environment, it is all talk and not that much action. It is something that has to change.

Two countries that have increased solar usage happen to be Germany and Japan. Spain, France, Italy and South Korea are next in the list and where is the United States? Well, one thing is for certain and that it is not in the top 10 despite the fact that it is an industrialized nation; however President Obama and his team are working on improving U.S position in world on solar energy.

Chapter 6: Solar Energy Collecting as an Alternative Energy Source

Photovoltaic cells; those black squares an array of which comprises a solar panel are getting more efficient, and gradually less expensive, all the time, thanks to ever-better designs which allow them to focus the gathered sunlight on a more and more concentrated point. The size of the cells is decreasing as their efficiency rises, meaning that each cell becomes cheaper to produce and at once more productive. As far as the aforementioned cost, the price of producing solar-generated energy per watt hour has come down to $2.00 at the time of this writing. Just 5 years ago, it was nearly double that cost.

Solar powered electricity generation is certainly good for the environment, as this alternative form of producing energy gives off absolutely zero emissions into the atmosphere and is merely utilizing one of the most naturally occurring of all things as its driver. Solar collection cells are becoming slowly but surely ever more practical for placing upon the rooftops of people's homes, and they are not a difficult system to use for heating one's home, creating hot water, or producing electricity. In the case of using the photovoltaic cells for hot water generation, the system works by having the water encased in the cells, where it is heated and then sent through your pipes.

Photovoltaic cells are becoming increasingly better at collecting sufficient radiation from the sun even on overcast or stormy days. One company in particular, Uni-Solar, has developed solar collection arrays for the home that work well on inclement days, by way of a technologically more advanced system that stores more energy at one time during sunlit days than previous or other arrays.

There is actually another solar power system available for use called the PV System. The PV System is connected to the nearest electrical grid; whenever there is an excess of solar energy being collected at a particular home, it is transferred to the grid for shared use and as a means of lowering the grid's dependence on the hydroelectrically-driven electricity production. Being connected to the PV System can keep your costs down as compared to full-fledged solar energy, while at once reducing pollution and taking pressure off the grid system. Some areas are designing centralized solar collection arrays for small towns or suburban communities.

Some big-name corporations have made it clear that they are also getting into the act of using solar power (a further indication that solar generated energy is becoming an economically viable alternative energy source). Google is putting in a 1.6 megawatt solar power generation plant on the roof of its corporate headquarters, while Wal Mart wants to put in an enormous 100 megawatt system of its own.

Nations such as Japan, Germany, UK, the United States, and Switzerland have been furthering the cause of solar energy production by providing government subsidies or by giving tax breaks to companies and individuals who agree to utilize solar power for generating their heat or electrical power. As technology advances and a greater storage of solar collection materials is made available, more and more private investors will see the value of investing in this "green" technology and further its implementation much more.

Chapter 7: Solar Energy to Supply the Needed Power

The world is currently suffering from energy and fuel crisis. Fossil fuels are incessantly depleting and so energy bills are soaring higher and higher. Many homeowners and establishments are now complaining about their huge expenses. But did you know that there is an answer to the crisis and it has long been available to man. However, because of its inefficiency, it is not recognized as a viable solution. Solar energy is free but the cost of generating it is a bit expensive. With the vast improvements in technology, there are now more affordable options for those who want to use solar energy to supply power in their homes and offices.

Today, you can find a lot of manufacturers that market solar power kits. How effective are these power kits anyway? Many households are now utilizing the kits because they are efficient and convenient in generating the needed power from sunlight.

Battery chargers that are powered by solar energy are now widely available. Now, you don't need to rely fully on conventional electricity to power electrical devices. If you install the right size solar panel, you may be able to decrease your dependence on conventional energy sources. The energy from the sun is free and you will simply need to invest on the kit for a good power system.

People usually think that without sunlight, power can't be generated. The power kits are not only designed to capture the sun's energy but also to store the energy as well. During sunny weather, energy from the sun is converted into electricity and those that are not consumed will be stored. The stored energy can now be used at night or during unpleasant weather.

With portable chargers, you won't be able to get enough power. You can only use them for recharging small electric devices and battery packs. Cellular/mobile phones can be charged using the portable solar power chargers. The power is easily drained out and so it's not a very efficient source of energy. Despite the shortcomings of these devices, it would help if you still have them around. Who knows... a few years from now, researchers and scientists may be able to make a major breakthrough in solar power.

Since the power kits are affordable, a lot of people are now using them. A solar panel usually has one charger where you will plug the equipment. The solar chargers are very useful and you can carry them even in your hiking trips. Just place them on your car's dashboard where sunlight is accessible.

The portable chargers or solar panels can be mounted in the rooftop to get a lot of sunlight. The large panels are still expensive and not many people can afford to buy one. However, governments can subsidize the investment to set up the power system. If you live in a place where the government offers subsidies, grants, and loans for the use of solar power, you are quite lucky. Take advantage of it so that you can now use free energy. Tax incentives are even offered in some places to encourage residents to go for solar power.

If you are still not sure about the efficiency of solar power, you can start by using a power kit. Try it now and expand the system later on.

Chapter 8: Solar Energy – New Discoveries

The day may not be far off when the use of solar energy becomes a norm. There is now a deep conviction among experts that given a few years time, solar power will be in high demand that the cost will go down, inexpensive enough to undercut the prices of oil-generated electricity.

Previous predictions that it will still happen in a decade may no longer be true. The anger generated by the recent prices in oil and its vulnerability to market forces and other events may have already been enough to polarized people, governments and scientific communities into seriously considering a reliable alternative energy source.

You cannot get a source more reliable than the sun. Even today homes that uses its power does not only benefit from the silent, energy generating, inexhaustible power of the sun, it also spikes up the prices of their homes. Those that have solar powered homes are even reimbursed for the surplus power that they supply to the power grid.

Presently, heliostats, photovoltaic cells and plate collectors are being used to collect the energy by focusing these panels towards the sun or constructing and installing the panel's on spots where the sun shines most. Development in technology as we all know often has a snowball effect. It never stops rediscovering and reinventing that the speed of development could often be surprisingly fast.

Today, a polymer foil, thin as a sheet of paper and lighter by 200 times when compared to the regular glass collecting plates, are being developed. Chances are, these new inventions and discoveries could very well have a great potential for mass production. Previously, the glass-based materials used for heat collection need expensive substrates and require additional support for mounting due to its weight. The polymer foil, being very light could now be attached even to the walls of a structure.

So confident are scientists in the development of this technology that while the polymer foil is being developed, a plastic solar cell, based on nano technology is gaining breakthroughs. This plastic material can collect the power of the sun even on a cloudy day through harnessing the infrared rays is believed to be five times more efficient than the current technology.

While plastic materials for harnessing the power of the sun are not new, it is only recently that this plastic composite could harvest the infrared portion. Previously, only the visible rays are generated, the infrared part, which is half of the power of the sun, is invisible.

Currently, the best plastic solar cells could only harness 6% of the suns energy, with further study and development, this new plastic solar cells are expected to harness 30% of the suns solar power.

Scientists and researchers alike agree that ultimately, solar farms will be harnessing all our energy requirements and costs of power will drop. Today the price of solar powered energy is about 3 to 4 times per kilowatt hour compared with conventional electricity. That could change dramatically though the development of the existing technology and recent discoveries.

The roller pressed flexible plastic materials and the polymer foil are only two of the best hopes in arriving at a cleaner, greener and safer environment that could ultimately free the planet from its dependency on the depleting supply of oil.

Chapter 9: The Cost of Solar Energy System

Solar energy is free but the tools that will enable you to harness this free energy are not.

If you are planning to install a solar energy system at your home and would like to know how much it would cost, there are 3 things you need to consider:

First, the cost depends on how much energy your household requires. More energy you require means more solar panels and higher cost of installation. The good thing is; the technology to harness the power of the sun is relatively cheaper than it was several years ago.

Second, the government is offering a rebate to those who will switch to alternative form of energy. That means; the government will shoulder a portion of the total cost of the system.

Third, you can get tax credits when you switch to solar energy system.

All these affect the overall cost of your installation.

A photovoltaic system can cost roughly at $8000 to $10000 per 1kW system. This equates to $8 to $10 per Watt. An average American household with 3 bedrooms requires at least 1.5kW to as much as 3kW. Thus, installation can cost from $14000 to $30000, before rebates and tax credits.

Government rebate varies from state to state. As a reference, an average California household that requires 1.5kW system can get as much as $4200 rebate (1.5kW= 1500 watts x $2.80 per watt = $4200).

A household that requires 3kW system can get as much as $8400 (3kW= 3000 watts x $2.80 per watt = $8400).

Before, tax credit goes as much as 30% but it is lowered to 7.5% of system cost after rebate.

So an average household with 1.5kW system can get $735 ($14000 - $4200 x 7.5% = $735) worth of tax credit while the 3kW system can get $1620 ($30000 - $8400 x 7.5% = $1620) worth of tax credit.

Example A: 1.5kW system Cost of solar electric system: $14000 Less rebate: $4200 Less tax credit: $735

Cost of the system after rebate and tax credit: $9065

Example B: 3kW system

Cost of solar electric system: $30000 Less rebate: $8400 Less tax credit: $1620

Cost of the system after rebate and tax credit: $19980 Again, these are rough estimates just to give you an idea of the cost of installing solar energy system. They do not represent real figures.

Top tips:

Energy independent home has higher resale value and is more preferred by home buyers. So, if you are planning to sell your home in several years, installing solar energy system does not only provide substantial saving on electric bill, it also is a good home improvement option.

If you want to finance the cost of your solar energy installation, include it in your mortgage.

If you want to reduce your electricity load, switch to energy efficient appliances. Buy electric appliances with ENERGY STAR seal.

Low energy requirement means lower installation cost.

Residents of areas with higher electricity rate can benefit most on solar energy system.

If you can't afford the photovoltaic system, you can install solar hot water system that costs roughly from $2000 to $4000.

Solar energy is best for houses that are located far from the existing power lines.

Chapter 10: Facts about Solar Energy and Solar Power Plants

The earth receives more than enough energy from the sun in an hour to supply the world's energy requirement for the whole year.

Unfortunately, only a tiny portion of it is harnessed and the world still relies on power plants that burn fossil fuels. The good thing, though, is that there is a constant increase in demand for solar energy; and over the years of continuous development, solar panels are much cheaper today.

During peak hours, the maximum power density that the sun can give is about 1kW per square meter. In other words, one square meter of solar panel can produce as much as 100 GWh (gigawatt hours) of electricity in one year. That is enough to power 50,000 houses.

If a solar power plant is build on 1% of the total land area of the Sahara desert, it will satisfy the world's energy requirement.

The efficiency of solar panels depends on several factors such as pollution, clouds, temperature and atmospheric humidity.

Solar power plants are very similar to other conventional power plants – with one significant difference: The majority of power plants draw their power from fossil fuels like oil, coal and gas.

When power plants burn fossil fuels, they produce greenhouse gases that contribute to global warming. Solar power plants or solar thermal power plants (or Concentrating Solar Power plants) utilize the power of the sun's rays to generate electricity.

The process could not be any simpler. The solar panels receive heat from the sun, which will be reflected to the receiver. The receiver converts into steam the concentrated solar energy. The steam is stored on tanks which will be used to turn the turbines and generate electricity.

The whole process does not involve any burning of any fossil fuels. Thus, solar power plants do not contribute to global warming.

The increase in the use of solar energy will bring down the demand for oil.

Today, there are more than 10,000 households with solar energy systems and the number is constantly increasing. If the demand for solar energy as well as other forms of alternative energy, the demand for oil will drop and the cost fuel will likely to follow.

Residential solar energy system can turn your electric meter backwards. Given that you are connected on a power-grid, the excess energy that your solar energy system produces will go to the electric lines to be used by other homes. As a result, any excess energy you give will be reflected on your bills. Your electric supplier will even pay for the electricity you supplied.

Residential solar energy system can save you money.

While the initial cash out for installing solar energy system at home is big, the device will pay for itself in the long run. Not only you will save money on solar energy system, you also help the environment by not contributing to carbon emissions.

Solar energy systems are reliable and can last for a very long time.

PV cells are last from 25 to 40 years. Many manufacturers of solar panels give 25 years product warranty. This is the assurance that solar panels are very dependable.

In addition, solar panels require little or no maintenance and the can be installed on most places where there is sunlight throughout the year.

Chapter 11: A Bright Future for Solar Energy: an Alternative Energy Source

I was first introduced to solar energy in the movie, Race the Sun with James Belushi and Halley Berry in the lead. It was a story about low- income and under achieving Hawaiian students encouraged by their teacher to join the Solar Car race. In the movie, a car shaped like a cockroach and covered with solar panels used the sun's rays as an alternative energy source to run the car.

Solar energy is the light and the heat from the sun. Solar energy is free and its supplies are unlimited. There are no air and water pollution caused about by using solar energy. But there are still some impacts on the environment although indirect.

Photovoltaic cells used to convert sunlight into electricity uses silicon and also produce some waste materials. There are also large solar thermal farms and these farms can also be harmful to the environment and desert ecosystems if not properly managed.

Solar energy can be used on different aspects. Solar energy can be used in agriculture. Greenhouses (which are entirely different from greenhouse gas) convert solar light to heat to be maximized in enhancing the growth of plants and crops. Greenhouses have been around since the Roman times and modern greenhouses were built in Europe in 16th century. Greenhouses are still an important part of horticulture nowadays,

Daylight systems are also being used to maximize the energy released by the sun. It is used to provide interior illumination replacing the artificial lighting. Daylight systems include sawtooth roofs, light shelf, skylights, and light tube. Daylight systems when they are properly implemented can reduce lighting-related energy consumption by 25 percent.

Solar energy can also be developed into solar thermal technologies which can be used for water heating, space heating, space cooling and process heat generation. Solar energy can also be used to distil water and make saline or brackish water potable or drinkable.

The solar water disinfection or SODIS involves exposing water-filled plastic polyethylene terephthalate or PET bottles. This process takes a long time, since the exposure time varies on the weather conditions. It requires a minimum of six hours to two days during days with overcast conditions. Currently, there are two million people in developing centuries use SODIS for their daily drinking water needs.

Also sunlight can be converted into electricity using photovoltaics or PV. PV has been mainly used to power small and medium-sized things like a calculator powered by a single solar cell. There are homes powered by photovoltaics. Using solar energy for water and space heating is the most widely use application of solar energy. While ventilation and solar air heating is also growing in popularity.

There are three main ways in using solar energy. The main way of using and converting solar energy is by using the solar cells. Solar cells convert light directly into electricity. Solar cells are also called photovoltaic or photoelectric cells.

Meanwhile, solar furnaces use a huge array of mirrors to concentrate on the Sun's energy into a small space and produced very high temperatures. Solar furnaces are also called 'solar cookers". A solar cooker can be used in hot countries to cook food.

With all the benefits if using solar energy, there is still a downside for this alternative energy source. It does not work during night time. The cost of setting up solar stations is expensive, but the benefit of using solar energy when accumulated is so much more.

Chapter 12: Solar Energy – the Future of Generating Energy for the Home

Solar energy for residential houses is nothing new. It has just been relegated to the background in lieu of rising cost of real estate; newer more advanced building materials, design and the limitation of resources.

Since man started building homes, sunlight played a major influence in the design. In fact, even in the more advanced urban planning method of the Ancient Chinese and Greeks, the orientation of the buildings is as much as possible directed towards where it could capture the most sunlight.

The ancients might not be as intellectually sophisticated then to use catch phrases as passive solar and thermal mass but when they build, they were building in compact proportion, employing overhangs, producing insulations and building in manners that direct the airflow within the structure and producing well lit, well ventilated spaces using the relative position of the sun to the orientation of their structures.

Lately, as the conventional sources of energy became more expensive, homeowners were once again turning to the sun for energy requirements.

Since the 1950's, harnessing the sun's rays has been developing and today the solar cell technology has achieved very efficient levels that modern (so-called green house) designs apply the sun's power to provide energy for the home.

While solar energy is free, the device that will convert it to run our appliances is not. To provide solar energy for the home, solar cells called photovoltaic made from semi-conducting materials, are grouped into modules. These solar panels are mounted on rooftops, yards or open spaces where it can capture the maximum amount of sunlight.

Whenever possible, the panels will be installed facing south to get the most out of the sunlight but tracking systems are also used to follow the direction of the sun. The solar panels collect the energy from the sunlight. The process basically is that when the panels are exposed to sunlight, the electrons are separated from the atoms. This movement of the electrons creates electricity.

To store power, pumps are often used - circulating water in the cells. The water goes into a storage tank where the power is stored, ready for use. Sometimes, the use of gravity is employed if it will just the same store the heated water in to the tank.

In spite of all the development in solar energy though, the use of this technology is not enough to provide power to the whole house. The best method so far can only fulfil about 80% of a households power needs. The employment of solar energy for the home will still require the use of the conventional power distribution method.

Powering the homes by solar means will still, for a while be augmented by a local power distribution agency. To many, this is already a good starting point. Homeowners that feel that the high cost of powering their houses through solar power, is justified when compared to the price that is now being paid for conventional electrification method where horrendous amounts of CO_2 are being dumped into the atmosphere just to generate a pitiful amount of electricity.

However, due in part to the rising costs of energy, the technology for solar energy has been undergoing rapid phases of development. Experts are confident that within five years, powering the home through the solar method will be made widely available for those who prefer it as its sole energy source.

Chapter 13: Using Solar Power for your Home Appliances

Solar energy only accounts for 1% of all electricity use in the United States. Though solar energy has been called a joke by other energy industries, with the renewing interest in alternative fuels it is becoming viable again. Solar energy is very expensive at this point but some people have found a way to make solar energy or partial solar energy part of their lifestyle.

Solar power can supplement your regular grid power by being attached to only certain electrical outlets in your home. The initial setup is expensive but if you want to keep your freezer or your refrigerator, which runs constantly, on solar power, you have an opportunity. Start small with a small solar panel at a time. You'd be amazed at how much money you can save by having just your hot water run on solar power.

You can break the bonds of the gas industry as propane and natural gas prices are starting to rise. If you have a natural gas or propane water heater, a simple solar panel can provide enough energy for you to have hot water for all of your family's hot water needs. This could be hygiene needs, cooking, your washer and dryer, or any other hot water uses.

If you are thinking about replacing your existing water heater look at an electric water heater instead.

Electricity off the grid may be expensive for a water heater but if you supplement with solar power the cost per month will be next to nothing. Find out how many watts per hour that appliance needs. Then contact a solar power specialist and find out what size of solar panel you need to produce enough energy for the appliance. You can install the solar panel yourself or have it professionally installed. Once the equipment is bought and installed you can also purchase a timer for your water heater.

A hot water heater timer allows you to set the time of day or night you want your hot water heater to come on. If you go to work at seven o'clock in the morning and take your shower around 6:30, the timer will come on about six o'clock and heat the water to your desired preference. The same thing goes at night. If you are washing dishes at a certain time or taking a nightly shower, the timer will be able to give you hot water when you need it. It takes a little getting used to this because most people want hot water on demand. Heating water is an expensive process and wastes energy.

While having your hot water heater totally on solar power, it may take a few years for you to recap the initial purchase price. Solar power at this point is a long-term investment. But once the initial cost of installation is set up you can run many appliances off solar power even that hot water heater that died and should have been replaced long ago.

As solar power and solar power panels are gaining in popularity and use, their prices should go down within the next decade. Keep your eyes open on the internet or while watching the news to see if there are new and better solar panels and solar cells being invented. Contact some companies; you may even be able to set your house up as an experimental station and have a solar panel for free to test the technology.

Try other alternative energy sources such as wind power to couple with your solar energy panels and double your output of energy with two renewable energy sources instead of one.

Chapter 14: Heating your Home with Solar Energy

It doesn't matter if you are building your home or remodelling, you can turn it into a solar energy home by making a few simple changes to your plan. If electric and gas become hard to manage you may want to consider heating your home with the sun. Solar energy is the heat that comes from the sun down to the earth. When it reaches the earth it spreads evenly but you may need it to go to a certain area like your home. How do you get that much sunlight to heat a home? It's easy to do and takes a few extra steps to help get it started.

Building or Remodelling your Home

If you are building your home you have several choices to choose from regarding your heating source. If you choose to heat from the sun you need to build your home facing in the direction that the sun rise's. This allows your home to get the most sunshine during the hottest part of the day. Buying solar powered glass windows allow the sun to come through and stay in the home without escaping back out. After the sun goes down your home is kept warm by the sunlight that came into the home during the day. You need to keep the door shut in order to keep the heat in and you also need to use insulated curtains on the windows at night so that the heat will not escape at night while you sleep. Make sure you don't allow too many windows on the side of the house that faces the evening sun as it may cause the home to cool down quickly.

Remodelling your home to use the sun as a natural heating source is fairly easy to do. Although you can't change the direction that your home is built in to face the morning sun you can still trap the sunlight that shines through and reduce the amount of time that you use another source of heat. You may want to consider building a sun room onto the side that catches the morning sun allowing it to heat up naturally and then install ceiling fans that will circulate the air into the parts of the house. During the day this may provide enough heat to maintain the warmth in your home. When remodelling your home, it will help to install solar power windows that are specially designed to attract the sunlight and allow it to come into the house but not let it escape. This is a natural way to heat your home.

Using sunlight to heat your home is an excellent way to save money on your heating bill and also to improve the environment. You can install a backup heating source in case the sunlight does not heat your home efficiently during the day because of clouds. Your back up system can be used to assist the solar energy which will also cut down on the use of electric or gas.

Chapter 15: Solar Energy – this Commodity is not for Sale

If anything good came out during the recent increases of the fuel oil prices, it is that once again, there is a merry interest in alternative sources of energy. Even when pump prices has been decreasing like nothing we have seen before, the uproar created and the pain it did to business will have a good chance of sticking. The desire for alternative sources of energy is on the forefront and may it stay there for good - as it should.

Extracting oil from crops is a good idea; the downside is that food supplies could be dramatically reduced. Wind power is another excellent thing except for the many buts that wind power generation have.

If costs is the main objection to solar power generation that should be the least of worries.

The installation of solar panels is until today, considered a specialized job. Like any commodity in the market, when the demand is high but the supply is limited, the cost increase. As more and more homes clamour for alternative sources of energy, better technology and more labour is drawn to the job that market forces could take place and result into much lowered prices. This though is still in the future.

Today the reality is the instability and the unpredictability of pump prices. But even if it does provide stable and predictable price movement, solar energy is free and it is inexhaustible.

While the technology of tapping out this resource is not, homes that have solar powered are getting back dividends in terms of higher appraisal for their homes, confidence in not being surprised by power shortage and outage, not being dependent on the fluctuation in power prices and the definite advantage of having provided a better environment solution.

Today, the typical ways at tapping this resource usually are:

1. Through a Heliostat – this are focusing collectors composed of mirrors that are aimed at the sun to collect the energy. The temperature that heliostats could provide reaches more that 4,000 degrees centigrade. This high temperature is sufficient for use even for furnaces.

2. Through Flat Plate Collectors – employ a system of pipes. The water inside the pipes becomes heated and is ideal for heating purposes like schools, homes swimming pools, offices etc.

3. Through Solar Distillation – instead of heat, this provides water. The mechanics is similar to the processes of plate collectors except that this is generally used to steam salt water. To do this, tanks and ducts are usually installed in surfaces that receive a good sun. Through the

heat, the salt water turns into steam and when the steam condenses, the water is collected for regular use.

4. Through photovoltaic Cells – These are the most common type of collecting solar energy characterized by solar panels installed in rooftops and other flat surfaces that there is a good sun, converting the power collected into electricity.

Unlike any products, whether refined or manufactured, the processing of solar energy do not need additional costs of energy to power it up. It is energy generating by itself. Except maybe for regular inspection and replacement of parts (when it employs a mechanical device), the tapping of solar energy is virtually maintenance free. Once installed, it could be used for as long as needed for the amount of energy, (depending on the capacity of the unit), which the owner requires.

Chapter 16: Solar Energy - a Sustainable Power is Harnessed

Since ancient times the sun was considered as a source of energy: spiritual and other wise. It is so sad to find out that in fact only 10% of solar energy is actually used. Perhaps if we are more aware of its use and its capacity as a sustainable power then maybe we will be able to use and promote solar power.

The Beginning of All Things

Unbeknownst to a lot of people, solar energy actually is the source of numerous sustainable powers like radiation, waves and wind. Solar energy has many usage it can give us light, heat, promote cooling, it can be harnessed through technology to power many things like for machines cooking, distillation, hot water, and disinfection.

Technology and the Sun

As we all know heat produces unfathomable amounts of energy. This energy has to be aided by technology in order to be converted into something that is usable by mankind. There are two types of solar technology, the passive and or the active solar energy. A classification between the two depends on how the heat from the sun is harnessed and channelled into ordinary things powered by electricity.

Active Solar Technologies use solar (photovoltaic) panels, combined with solar thermal collectors, and then channelled thru mechanical or electrical equipment. Passive solar technology is merely a technique in order to capture the suns useful rays; for example a skylight.

Electricity From The Sun

We are all familiar with the term 'solar panel"; solar panels convert the heat from the sun into actual electrical current with the use of what is called the photoelectric effect. Concentrated solar power produces insurmountable energy. In fact it was greatly utilized during the time of the Ancient Chinese Civilization.

To concentrate the power of the sun, a series of mirrors and lenses are used in order to focus the light in one area thereby producing a single beam. There are a lot of technological advances out there that concentrates solar energy in order to produce a concentrated amount of power they are; the solar power tower, the parabolic dish and the solar trough.

Dilemma, Solution, and Economics:

The primary concern of the use of solar energy as a sustainable power is that there is no sun during the night. Modern times require continuous supply that is why storing solar energy is a key component of solar technology. Thermal storage systems can store solar energy. Newer scientific discoveries also paved the way for thermal mass storage systems which vary storage capacity and function by storing more energy during off peak times and varying supply at peak consumption hours.

It usually takes a crisis for people to actually look at possibilities, and like with almost everything else solar energy began getting attention after the 1979 oil crisis and the 1973 oil embargo. In fact solar technology began its appearance in the 18 hundreds. IN the past solar energy as a sustainable power was a dream but cannot be realized due to expensive technology needed to use it.

Today, with the rising costs of electricity, the volatile oil prices, and its degenerating source have paved the way for solar energy once more. The consciousness environmental welfare has prompted companies to manufacture affordable solar technologies and sell it commercial. Indeed it is a fact that solar technology is expensive but it is only in the purchase of the machines needed in the long run if you calculate it, you are actually getting a bargain.

Chapter 17: Pros and Cons of Residential Solar Energy System

Harnessing the power of the sun's ray to create energy to power our house is very appealing. But the question is, "Is everything about solar energy good?"

Looking at the current price of fossil fuel-based electricity, it is quite impractical to convert into solar energy system. However, with the growing concern on the state of the earth, there is really a need to find other means of energy aside from what power plants are using right now. Where do you place yourself?

Whether you are an advocate of clean energy or simply care about where your finances go, looking at the pros and cons of residential solar energy system will help you decide on whether to convert or not.

Pros

Solar energy is free. Did you know that the earth absorbs 174 pettawatts of solar radiation? This means that we have more than enough source of free energy to power every house in the world. Unfortunately, most of our energy is still drawn from oil, gas and coal. But in recent years, there is a steady increase of demand for alternative and renewable energy like solar power. It is estimated that the demand for alternative sources of energy will increase by 53% between 2011 and 2030.

Solar energy is clean, renewable and sustainable. Because the energy created from the sun's rays does not produce by-products like those from fossil fuel power plants (sulphur dioxide, nitrogen oxide, mercury or carbon dioxide), it does not contribute to pollution. Accordingly, the increase in the use of solar energy and other alternative forms of energy will decrease the demand for greenhouse gases-producing power plants.

The price of photovoltaic cells is steadily decreasing. The demand for solar panels has risen by 57% in the United States in 2010 and is steadily increasing on a monthly basis. The increase in demand results to the improvement of solar technology as a whole. The prices of photovoltaic cells have declined on the average of 4% every year over the past 15 years.

Solar panels can be installed on most rooftops, eliminating the problem of finding a suitable place for installation. Solar panels require little or no maintenance. The original photovoltaic cells technology is used for most satellites orbiting our earth today which are not maintained at all. Many solar panel manufacturers give 25 to 40 years warranty on their products.

Because most areas of the country receive a substantial amount of sunlight throughout the year, solar panels can be installed anywhere.

Many states and counties in the United State and UK give tax credits and rebates to households who want to install solar energy system. Check with your state or county government the cost of these incentives.

Cons

While the prices of PV cells are in constant decline, the cost of installation is substantially high compared to the current electric cost. But the good thing is, after your initial cash out, you don't have to pay every month on electric bills for the rest of your life.

On areas cities and areas with heavy pollution problem, solar energy may not work as fine. Weather can also affect the efficiency of solar energy. If it is raining, overcast weather or if there is a hurricane, the solar panels' efficiency is decreased.

You are only producing energy during day time.

These are general pros and cons you might encounter when considering the conversion to solar energy system. It would be best if your decision is based on location, cost, budget, rebates, tax credits and practicality.

Chapter 18: You can Build your Car Powered by Solar, a Green Energy Source

All cars are powered by petrol and diesel. But with the volatility of crude oil prices and because it is not a renewable, something must be done before it is too late. Given that solar energy, a green energy source is used to power a community, you can also do this on a small scale by using the same principles to build your own car.

But what do you need to make this work? A lot of things but the two most important are the solar array and the batteries.

The solar array is vital because this is what's used to collect the sun's rays and then converts this into electrical energy. There are two types to choose from in the market namely the prefabricated type and the individual kind which you set up yourself.

The lower end model which can produce a significant amount of power is the terrestrial grade version.

Proper wiring must be done to make sure that if one of the panels is not working, your vehicle will still move. If you are worried that the voltage of the solar array should match the system voltage of your motor, you should not worry because it will still run.

We mentioned earlier that the battery is also important because this is where the solar energy will be stored. Your options for this are lead acid, lithium-ion or nickel-cadmium. Just how many you need to buy will depend on your motor's voltage.

When you finally have these two components, these will now have to be connected to the motor. So you know how much juice is left in your batteries, you will also need to install instrumentation similar to the heads up display console on regular cars which tells you your speed, mileage and gas.

Don't forget to put a steering wheel, suspension, brakes, tires and hubs. You may not be able to make a car that have the same features as like what you see done by one of the three US automakers but just enough to be able to drive it from one place to the next.

The only cars that use solar energy so far are the ones only used in races especially the one held in Australia from travels from the northern part of the country all the way to the south. If this has helped people realize that renewable energy is really the key to the future, the big automakers should try tapping this technology instead of relying of still relying on gasoline.

But apart from solar energy as a green energy source, biodiesel is another alternative. This is a combination of alcohol like methanol and a chemical process that separates glycerine and methyl esters (biodiesel) from fats or vegetable oils. This can also be done using corn and sugarcane.

Although these are not renewable, these are still considered as a green energy sources because it is cleaner than conventional gasoline. This means you do not release harmful chemicals such as carbon monoxide into the air which causes damage to the environment. So if you can't build a solar powered car, consider a different fuel alternative.

Chapter 19: Solar Panels for Today

One of the main problems with solar power is collecting the power of the sun and at the same time having the solar panels pleasing to the eyes of the people around you. You have probably seen solar houses in your neighbourhood with five or six solar panels reflecting the sun. Though this is the wave of the future, it is not at this time pleasing to the eye of most consumers. These home owners put their solar panels either on the roof or in the backyard and to the casual non echnologically-oriented person this looks tacky and takes away from the beauty of a neighbourhood.

There is new technology that allows the solar power user to set their solar panels up in a way that is aesthetically pleasing, and at the same time gives greater opportunities to collect sunlight. These designs have been put in place to serve dual purposes. They not only enhance the look of a house, but also give the owner a chance to use his solar panels as cover for other things. Solar panels can be set up in a variety of ways as long as they are angled at the sun and collect the optimal amount of energy that is emitted in that area of the house.

These new designs have combined traditional house extensions with a revolutionary new concept that allows the solar panel to act both as a part of the house and at the same time collects energy. One such design looks like a carport from the outside.

The solar panels are set up so that they extend from the house above the car parking area. The driver of the car can park his car underneath the solar panels and they, in turn, give the car protection from the elements. Snow and rain cannot hit the car and the solid construction of the solar panels helps them support themselves, even under the weight of snow. These solar panels have a heating system that goes throughout them. The heat will melt any moisture such as snow or ice so that the sun can get through to the solar panels no matter what the weather.

Some designs have solar panels sloped gently from the second story of the house to provide an area for which a person could put a garden box or flowers and have some protection from the sun during inclement weather. Other systems include panels that will come out during sunrise and can be retracted when the sun goes down.

Another technology advance allows a greenhouse to use only solar power. This will allow the heat to be trapped inside the greenhouse causing the plants to thrive. At the same time the sun collected above the greenhouse can be reflected down to the plants to give them the solar energy they need to live. This solar panel can only collect so much energy within each cell then the residual sunlight would go to the plants. The added benefit is that the installation of these solar panels will cause the humidity to be collected inside the hothouse and allow for the house to be warmer when the weather cools down. All this put together does not take away from the idea that the house is receiving renewable energy that cost nothing except for the setup of the solar panels in the first place.

Check out the Internet for the new designs of solar panels. These solar panels are becoming more attractive than is aesthetically pleasing to the eye because they are becoming more efficient to help you save money and to be less dependent on fossil fuels.

Chapter 20: How to use Solar Power for Cameras

If you are a professional photographer or you are just an amateur shutter bug, you know how frustrating it is to have your camera lose power during a vacation or an outing. It is just frustrating to be somewhere like Disney World, seeing a national park or on top of the Great Wall in China. In the middle of photographing your family by the very thing that you drove hundreds of miles to see, your camp camera battery dies. Some cameras have backup energy systems and last longer, but most batteries will die only after a few pictures.

There are rechargeable batteries which you can purchase but even if you use the batteries in your camera and then use the batteries that came extra with your charger, you still have only a limited life to your camera. The solution to this is solar powered cameras. You can take a small sheet of solar cells and place your camera in the sun and solar cells will recharge your batteries. This will allow you to be able to recharge and take the pictures you want to while at the same time you're not wasting electricity.

When you plug that charger into the wall of your home, you are paying for the electricity that is coming through. It may be a small amount but a lot of recharges can add up after a few years. The batteries, even though rechargeable, become old and do not work as well. You will be forced to buy new rechargeable batteries and throw the old ones away.

Batteries are a horrible addition to any landfill. As the outside casing breaks down, the inner core and acids drip through the landfill and can contaminate groundwater.

By using solar power to recharge your camera's batteries you are not only saving the environment by keeping batteries from going to the landfill, but you also save electricity. The sun will recharge those batteries and in return you will get a longer life from your camera during the time you're taking pictures. No more having to worry about running to the hotel room to grab the extra batteries or digging through your purse to find that extra battery pack to go into your camera.

Solar power is being used in more and smaller electronic devices. The camera, with its battery problems, has been the newest addition to the solar power equipment family. You can find solar power battery chargers at camera stores, your local camping, or outdoor equipment store, or they can be found on the internet.

Again some of these rechargeable solar panels are a little expensive at first but you can use them for different purposes other than your camera. Buy a solar panel that has a battery pack that can go with your rechargeable battery charger. If you do not need your batteries recharged, you can use the battery pack as a source of electricity for other light devices or a source to plug them into another device.

So next time you are on a vacation and you are about to take a picture of the Grand Canyon or a picture of Old Faithful gushing in Yellowstone National Park, you can have the confidence to know that your batteries are charged and that you did not pay for electricity that you are using.

Display the solar panels out in the sun, sit back, and relax. A good time to do this is when your family is eating lunch or dinner or when you are inside. Leave solar panels out in the sun and click away. You will never have a vacation or an outing without a chance to take pictures again.

Chapter 21: Solar Powered MP3 Players

Everybody seems to have a personal music device these days. Whether they have an iPod or a CD player, people from all ages and all walks of life want their music with them wherever they go. The negative aspect of having your personal music player or iPod with you is that sometimes your battery will run out and there is no place to plug in or recharge. Some iPods are charged only through a computer system. It is disappointing to be in the middle of your job or be in your car travelling somewhere and all of a sudden your iPod or music device goes out.

With the new micro solar powered technology that has recently become available, you can now buy a solar powered iPod charger. This charger becomes useful to the user as a backup power source which can be powered by the sun while you are listening or you can put your iPod charger in the sun to have that energy available anytime you wish to use it. It doesn't matter what iPod you are using or what generation it was made. The iPod solar powered charger will charge up any generation of iPod. This means that you do not have to search for a computer to charge with or hook into any electric power support.

It may not seem that you are saving that much energy by having a solar powered iPod charger. The energy that you save when added up with everyone else in the world having an iPod, which uses the same system, can save hundreds of watts of energy that would otherwise be generated from fossil fuels. Technology is great but with solar powered technology, you are more green-friendly and are able to protect the environment from dangerous pollutants and poisonous toxins.

The new solar powered iPod charger even works on the new iPhone. It is small and inexpensive. This works well on camping trips or long excursions where a power supply is not available. It is great to know that when you go on a hike or out on a boat you have your iPod powered all day long without losing the enjoyment of the music you are listening to.

It is very easy to operate. You simply open up the solar panel, which includes solar cells. It looks like a small sheet of aluminium foil. It will catch the sun's rays and the sun will convert its power into electrical power. The charger will store the energy and be ready for use whenever you run out of power in your battery. Simply hook up to the solar panels directly or use the charger to charge up your battery. Nothing could be that easy.

Solar power is coming of age. Although it is expensive to power large structures and buildings with solar power, technology is catching up by using solar power to get energy to smaller devices. To reduce global warming it takes baby steps. By using solar power for small devices you will be taking baby steps toward a cleaner tomorrow and a better world for your children.

There will be no need for disappointment anymore with your new solar powered iPod charger. You will be able to have your music with you when ever and where ever you want. You will no longer have to rely on computers or outside sources. This is simply just a better method of obtaining energy than the traditional method which includes looking into the power grid. The power grid costs money but with solar energy you do not have to worry about being harnessed to the power grid again and
because power is free.

Chapter 22: Personal Solar Energy Packs

A new invention by Solar Time Solardyne has brought new innovations to the personal utility solar power pack and has revolutionized the idea. You can use this portable power supply for camping, field research, emergency home power, or any of the things that you think of that would be useful in the field. It can even be useful as disaster relief. It is great because you are able to carry the power with you to a disaster zone if you are a first responder. Even missionaries have found it useful in a field where they can take electricity with them into areas that are void of power.

This personal solar power pack includes a folding 20/21 high-efficiency solar power panel. These solar cells are built with a mono-crystalline material that will produce more power personnel than any other solar panel on the market. It has a battery charged controller and state of charge indicators that will allow you to see how much power is in the unit and how much you need to meet the threshold. With a 300 W AC inverter using only one AC plug, you are able to patch into most electrical devices, lights, or camping appliances. The panel comes with jumper cables, a controller and even an extra port that will allow you to alternate your electrical needs as the situation dictates

The good thing about any personal backpack type solar power unit is that you can take it into almost any situation and be able to have power at your disposal.

Fire fighters jumping into a hot zone have carried these to be able to charge radios, flashlights, and other electrical equipment that will allow them to do their job more safely and efficiently.

Personalized solar backpack units can also be used for personal and recreational activities. In some instances, there are always times and places that you need a source of power for entertainment devices, camping equipment, or other necessary items in your personal toolbox.

A solar backpack unit like this one will allow you to enjoy being outside more and to be able to go further into the wilderness without worrying about your power needs.

Personalized solar backpacks are a convenient way for you to carry power around with you for use anytime you need it, whether you are going to the beach or just out of town. If you are a professional who works out in the field, such as a surveyor, field biologist, or any other person that needs power in a remote area, personalized solar backpacks will give you enough energy to get the job done.

For example, imagine being a worker in the middle of an important project and the generator is down. With a solar backpack unit, you would be able to have emergency equipment or even equipment that might just give you the comforts of home. All you have to do is have the battery charged in the sun. When you are ready to use the battery, you can plug in any AC or other type direct current device to receive the power you need to work properly in any given situation.

You can find personalized solar power backpacks on the internet or you can contact the company directly to see if these products might be right for you. They are small, lightweight, and portable enough for you to carry on your back to give you the electrical power you need when you need it.

Chapter 23: Outdoor Solar Powered Equipment

Solar power can be used for other things than just getting off grid electricity for your house. Solar power can be used in small appliances outside the house. If you are going camping there is nothing so bothersome than to have a small appliance such as a radio, thermos, cooler, or art and entertainment device go down because the batteries are dead. With solar powered camping equipment you can have all the luxuries of home and not have to spend a fortune on batteries that go dead after a few uses.

Solar powered camping equipment can be implemented into a number of devices. A flashlight or lantern is standard camping equipment. If you have a solar powered flashlight you can put the flashlight out during the day to collect the sun's rays and use the flashlight at night when needed. An average flashlight will burn for nine hours on two D-cell batteries. With a solar powered flashlight you do not have to replace those batteries. You will also save on landfill pollution. Without the thousands and thousands of batteries in our landfills, it would reduce the chance of chemical contamination of groundwater supplies.

Another piece of camping equipment that you can use at the campsite is an electric stove. Traditional stones and grills used oil, propane, or other fossil fuel based additives that not only cost money but add damaging pollutants in noxious fumes to the natural habitat in which you are camping. By using a solar powered stove you will catch the energy of the sun during the day and then when you turn on the stove at night the electricity stored in the battery will ignite and run your heating coils.

There are no pollutants or emissions whatsoever. This heat comes from the sun and is a clean and renewable source of energy.

When you go camping you may want to bring a small entertainment device like a DVD player, a radio, a CD player, and sometimes even a small television. You still use solar power on these devices by purchasing a solar panel battery combination in which all the energy that you would need would be stored into the battery as the solar panels collected from the sun. This will allow you to bring more electronics equipment to the camping site without having to rely on generators that spew fumes and gases into the air or your car which has to be run from electricity to be generated.

By using solar powered camping equipment you not only protect the environment, you are also being green friendly. You can bring more devices with you and make your camping experience all the more enjoyable. Some people do not like camping because it takes them away from the comfort of their home and the things they like to do is where electricity is important. Here you can take that couch potato, get them out in the woods where they can still enjoy the comforts of home, and at the same time see nature's wonder.

The initial cost of solar powered camping equipment is a little expensive but if you factor in the cost of batteries, the comfort of having your electronics devices with you, and the fact that you protect the environment. You will see that a solar powered camping trip is the best thing not only for you and your family but also for the environment and the world at large. You can find solar powered camping equipment online or at your local camping or outdoor equipment store.

Chapter 24: Using Solar Power in Boats

Have you ever been fishing or sunbathing on a boat? You probably wished you had something to listen to. Some people bring radios or CD players out to their boating excursions, but when the batteries are gone, they are gone. You could bring extra batteries but time and weather may have made a decrease in their potency and you would just be wasting time. That is why a marine solar powered battery is the best addition to any boat.

You can find these at boat shops, nautical parts stores, or you can order them online. It is amazing how much you can get out of these. There is not much of a supply of electricity on a boat. This is especially true if you have a john boat or just a little pleasure boat. You have the battery of a major engine but that's usually just a basic car battery which is hard to get the ampage and wattage correct for you to run your electrical devices upon them. With a solar powered marine battery you will have a source of electricity for all your electrical devices and even better, you don't pay a cent for it. The sun will hit the solar panels and charge the battery. When you need the battery you can plug in to a camping stove, flashlight, or even recharge batteries in the electrical equipment that you already have.

A solar powered marine battery is also good for emergencies. You may have not used your flashlight in a while but it is stationed upon your boat. The flashlight batteries could have been damaged by water or have just naturally died out. By using this solar powered marine battery you can hook up a light source to it to help you out in an emergency. It's a scary thing to be out on the water at night without lights. Other boats may run into you or there are a hundred different ways you can get yourself in really bad trouble. Having a light on board will signal to the other boats travelling in an area that you are actually there. You'll also be able to see obstacles in the water as you try to get back to shore.

Outside of emergency uses, the marine solar powered battery can charge the existing batteries that you already have. With the right equipment you are able to recharge your flashlight batteries, recharge the batteries in the radio, or the solar power can become your direct source of energy for that device. If you own a larger boat and it has appliances or other equipment that needs logical power, you can use the marine solar power charger to run some of this equipment.

Electricity on a boat is not cheap. The engine turns a generator which creates electricity. The engine runs on fuel and fuel is expensive. The engine also creates pollutants not only in the air but also in the water. Have you ever seen the oil trail behind a boat as it goes through the channel or across the lake? This is oil that is being polluted into the water.

By the use of solar power you can reduce the amount of pollution and have clean, free, green friendly power created right on your boat. A boat is a great place for solar power because there are very few shadows and you are out in the open. Lay your solar panels on the deck that will give them the most sun and go about your business having fun on the water.

The solar power will power up the battery or charger and you will have ample electricity to run your appliances, electronic devices, and electronic nautical equipment. Though a marine solar power generator or charger is a convenience it is almost an essential piece of boating equipment.

Chapter 25: What are Solar Ponds

Another technique similar to desalination pits is a solar pond. Solar ponds are not very deep. They usually run about three to six feet deep. The water in the pond is salt water. Salt water lakes and ponds have increased temperature. As you go deeper, the salt creates a density gradient. A density gradient will reverse or stop the convection currents that usually cool and heat water. During its experimentation phase, the solar pond produced 90° Centigrade temperatures. Not only was this is a giant in the amount of heat generated by the sun and the salt water, but the solar pond also produced 2% in solar to electric efficiency ratings.

Solar ponds used today have been very successful. At the University of Texas at El Paso, a solar pond heats the entire university. In other places where there are salt ponds, such as Israel or Utah, the salt gradient is used to produce heat for heating homes and other buildings in the community. The solar ponds have a dual use in which the salt and other minerals are collected after the water evaporates. By having a shallow salt pull, not only will you have solar produced energy, but you also have minerals and salt left behind at the bottom once the water evaporates. These can be used in a variety of industries. Sea salt and other spices are derived from the salt that is mined from solar salt ponds.

Some solar ponds have been used to grow certain species of salt water fish. In warmer waters, fish can be bred for food. For a non industrialized country that has salt ponds, one salt pond can produce enough energy and food for an entire village. Their extra income is also increased by the mining of the salts and minerals left behind.

A campaign to increase salt pond usage has been spreading through the third world countries by organizations such as the Peace Corps. Peace Corps volunteers and scientists have taught entire villages how to create a salt pond, use the water for heating purposes, and to grow aquatic life within the salt ponds to help feed the population.

Chapter 26: Solar Power in Space

Japan is joining the world's technology push for solar power to have a viable, profitable system readily available in the future. In the year 2040, Japan has plans to have a gigantic solar collector and generator put into orbit around the Earth. This sunlight collector and generator will be able to collect energy from the sun 24 hours a day. On Earth, the clouds and the rotation of the Earth only allow a certain amount of solar energy to be collected each day. In space, the solar power can be collected 24 hours a day without the obstruction of the atmosphere or the rotation of the earth.

The satellite will be able to generate over 1,000,000 kW per second. This is about as much energy as a nuclear power plant puts out. Nuclear plants have the reputation of leading to exposure of toxic radiation. Their use and the resulting danger to the environment have been documented with incidents like Three Mile Island and Chernobyl. Though improvements have been made to the nuclear energy field, an environmentally friendly power plant like the one that is proposed by Japan would make nuclear energy obsolete.

The satellite will travel about 36,000 km or 22,320 miles above the Earth's surface. With this geostationary orbit, this satellite will able to be gather power transmissions to an antenna that is over 1000 m in diameter. The electricity produced will be sent back to the Earth in the form of low intensity power comparable to those that are emitted by mobile phones. This power would be in the form of microwaves which would not interfere with any mobile phone service or other equipment on the Earth.

Solar power would be transmitted in such a way that it would take all the solar power that was collected within the dishes and transmit a concentrated beam towards receptors on Earth. The receiving antenna on Earth would be several miles in diameter. It would probably be set up at sea or in a remote desert area so that there would be no accidental interruption of telecommunication services. The satellite would weigh about 20,000 tons and the Japanese are looking at a cost of around $17 billion. This does not take into account the inflation that will rise in the cost of materials between the year 2040 and now.

Satellites such as this will be able to produce a clean and renewable energy source able to power the same amount of recipients as a nuclear power plant. It would also not leak toxic or poisonous wastes and the environment would benefit from the lack of greenhouse gases produced by fossil fuel power plants.

The world should come together with systems such as this and use the technology already in use for fossil fuels to power structures and infrastructures. This would greatly reduce the amount of greenhouse gases to such a limit that the effects of global warming could be reversed. It has been suggested that the United States and China contribute not only technologically, but also monetarily to the project to help the Japanese get off the ground faster. The efforts of the international community to place these type of orbiting power collectors into space would accelerate the ongoing battle against fossil fuel dependency.

Chapter 27: How the Government is Using Solar Power

While the debate goes on about how solar energy can save you money and how solar energy is not a viable resource. The Department of Energy has released details that they are funding efforts to concentrate totally on solar power. The mission of this effort is to see if solar power is as efficient as the solar power industries claim it to be.

Their ultimate goal is to see if solar power can bring the cost per kilowatt hour down at least $0.10. This sounds like it isn't very much of a reduction, but look at the billions a kilowatt hour spent every day in the United States and you can see the huge savings in both money and energy.

The governmental effort at concentrated solar power will use an array of mirrors to focus sunlight onto a central receiver. Once sunlight is received, fluid within the receiver is heated and forced through a turbine. Currently there are such efforts going on but the costs to build the technology for the heating of the fluid and a turbine generator is not going so well. The $5.2 million project is part of George Bush's Solar America Initiative. Ironically, George Bush was a baby of the oil boom and his push toward new alternative fuel technology might be a reflection of the political climate in America. He wants solar power to be commercially viable by the year 2015. That will not happen, however President Obama is trying to push this project through.

The Department of Energy has concentrated its efforts to collect solar research in which the cells turn sunlight directly into electricity. This is the debate going on in many think tanks in public and private arenas. Twelve projects are being funded by the grant under the Solar America Initiative. There will be nine separate private companies involved in the coordinated efforts, needing funding of more than $2 million.

The Solar Car Eight of Lakewood, Colorado will be looking at the development and advances to a system of collecting solar power. They have worked in the past using liquid salts to directly transfer energy, as with solar ponds. They are excited at the prospect of being able to build a better poly-meric reflector Other companies that are contributing to this project are 3M, Alcoa, Brayton Energy, Hamilton Sundstrand, Infinium, Sky Feel, and Solar Millennium. With the declining American dollar under a recession, the race to find alternative energy and products is accelerated. With the instability in Tunisia, Egypt and other gulf counties and oil prices soaring beyond belief, multinational efforts will be easier to gather once the United States's solar power initiative is at its greatest extent.

With projects like the Japanese solar satellite system, which collects solar energy through huge panels placed in orbit around Earth and then transmitted through microwaves to huge collector dishes, alternative energy sources such as electrical power, coal, gas, and propane could soon be a reference in the history books. The government is finally looking at a hard-line approach of dealing with alternative energy solutions and diminishing the need for foreign oil and fossil fuels. As the public sees the prices of gas rise while the costs of solar powered technology decreases, the public will soon follow this new technology to create a more green friendly planet and healthier environment for its population.

Chapter 28: Solar Power and Health

When a person decides to install solar power into his home or into the appliances of his daily life, he may think that he is living the natural lifestyle. Not only is he protecting the environment and also reducing dependency on the power grid and the need for foreign oil, but he is also saving money. If you want to go to solar power for a naturalistic approach, you may be surprised that solar power cells using blueberries instead of silicon in their makeup are currently being tested.

Blueberries are a healthy food and have all kinds of holistic health properties. They have been known to help a variety of health problems. Through technology they may be helping our energy problems also. Researchers at Tufts University analyzed more than 50 fruits for their antitoxins and capability. Amazingly, blueberries came at the top of that list. So now the blueberry is not only a health benefit, but researchers in Rome has announced that they have a new type of solar panel made with the pigments of the blueberries in which we eat every day.

They propose that they can make a solar panel that has no silicon whatsoever in it. This will make solar panels cheaper and more efficient. Since it is made with a natural material instead of silicon, the solar panel will be more flexible. Researchers contend that they can have a solar panel so thin that it could be transparent and still wrap around any solid object.

The production of solar panels and the cost of silicon have kept the technology of solar power expensive. It is believed that with a renewable organic compound made from the pigment of blueberries, the cost of creating a solar panel could decrease as much as 15% or more. This would create more electronic currents which the user could utilize. The research team in Rome has decided to develop the pigment solar panel more thoroughly before releasing the technology to solar power companies for further research and implementation.

This would make for an entirely natural process. Not only would you be using a renewable energy from a plant that can be grown just for the need of energy, but can also be used as an antioxidant and a healthy part of any diet. The transparent solar panel that can be created from the blueberry's pigment could be applied to the sides of houses, on the rooftops, on the sides of solar powered cars, and even solar powered boats. The uses are endless with a transparent solar panel. No more will you see the bulky metal frame solar powered panels jetting from the rooftops of houses or littering the backyards with unsightly clutter.

If you are thinking about converting your house over to solar power, you might want to really research the use of blueberry pigment solar panels. Check and see if the technology tool will be available to the public in the near future. You can save lots of money and also be able to make your house more aesthetically pleasing to the eye.

There is no limit to the places you can put a flexible solar panel. They can be fastened on the side of a shed, the copier doghouse, or just wrapped around your Clothes-line support poles.

The possibilities are endless with a little creativity and imagination. Keep an eye on the Web for more developments in this new technology and you might even want to invest in a company that's coming out with it!

Chapter 29: How to use Wind and Solar Power as Energy Sources

When you think about solar power, you think of a way to convert the sun's energy to totally heat and power your home. This can be done with great expense. If you want to totally live "off the grid" as it's called when you do not have to use the local power or fossil fuel plants. You would have to construct very large solar panels and be able to regulate the power going into your home. It's possible but there are other alternatives.

While a minor solar power project can reduce your power bill, you may want to think of going off the grid by using other alternative energy sources that combine with your solar power. Wind generation is a great way to add wind and sun together to make your power bills lower and be less dependent on the grid. By combining the two alternative energy sources you will use less electricity and, in the long run, be saving the planet by reducing greenhouse gases that are produced by fossil fuel energy producers.

A wind generator is a generator that is hooked to a fan that is turned by the wind. When the fan turns, the turbine creates electricity which is then converted and moved into your home. Combined with a solar powered panel, which collects the power from the sun and converts it to electricity, you can use this to reduce your need for outside power sources. The negative thing about either system is on a day without wind or a day without sunlight. On those days you would have to go back to relying on either the battery power that was saved during the days with sun and wind or you would have to rely on the power grid.

In the United States, the best places for solar and wind power are in the Great Plains area. States like Colorado, Wyoming, Nevada, Nebraska, and Kansas have a high wind factor but also receive lots of sun. You can place your wind turbine in the back of your yard or in an area on your property that receives plenty of wind. They should not be placed next to a tree or building, but somewhere a little bit away from these types of structures so that the wind can hit the turbine at full force.

You would want a rotating coupling on your wind-generator so that as the wind changes directions, so will the propellers that generate the electricity. The solar panel should be placed in an area that will get the most sun. Your house may be shaded with trees or other buildings, but there will probably be one spot in your yard or on your property that will allow for maximum sun exposure.

You can see that this is a great way to bring down your power bill and your reliance on the power grid, except for those cloudy or windless days when you will have to rely on the power grid system.

The long, winter storms can cover your solar cells, while the cloudy skies will not allow the solar energy to reach the cells as they would on a bright, sunny day. The wind could cease for two to three days and, again, you would be forced to go back to the power grid. The only way to avoid this problem is to have a full set of power collectors in which you could switch from the energy source to the power collectors and stay off the power grid for a little longer.

Chapter 30: Using Solar Power for Fresh Water

There are many parts of the world that have no clean water source or water in that area they can afford. Many Third World countries have trouble keeping their water clean from diseases and impurities. Solar power can help struggling nations by using solar desalination and solar disinfection. With the use of solar power, these communities can produce drinkable water and water clean enough to use in household activities. They are able to take a bath, to use the water to clean and disinfect, as well as water with which to cook.

Pooling water often will be used to desalinate or disinfect the water. The negative thing about doing this is that you use energy. You have to use wood to build a fire, use propane or gas, or use electricity to heat the water up to make it pure. The people in the Third World countries do not have the luxuries to do this. To use fuel is to use money and money is not something that these communities have a lot of. Using solar power to desalinate or to disinfect the water is a lot easier and does not use power, except from that of the sun.

To desalinate water, a solar powered desalination pit could be used. The concept is quite simple. A container of impure water is put into a pit. A piece of plastic is then stretched across the pit with a weight placed in the centre. The weight will cause plastic to bend toward the container in a cone shaped. As the sun heats the water, the water boils at a rapid rate and condensed through the plastic. Here the water can be collected and used. The water will be free from salt or from any other type of debris that might cause sickness or death.

Desalination can be accomplished in large-scale. If enough pits are used, this method could supply a needy community's clean fresh water. It is not the depth of the container that is important, but its width. You want a container deep enough to collect enough water to be able to go through the process in a day and evaporate, but shallow enough whereas all the water through will evaporate before that day is over. This may take some trial and error to make sure you get the most water on a given day. The container should be put in the pit deep enough to allow the condensation to take place, but not deep enough where the movement of the sun will cast a shadow over the desalination pit.

This method used by many countries in desert areas. It is solar power technology that anyone can use. The user, however, must make sure that the plastic is clean to start out with and that the condensation is collected quickly. The condensation could be a breeding ground for insect larva and other impurities that might land upon the plastic or on the water on top of the plastic. The ideal place for a desalination pit is usually in an open space away from trees or other objects, such as buildings that might shade the pit.

Also, the builder of the desalination pit should make sure there are no animals, such as livestock or domestic pets that will run through the pit and contaminate the water. If the pits are put into a dry climate, animals will most likely be attracted to water and sewer.

Crawling or flying animals like spiders, flies, and mosquitoes need to be monitored, making certain that they do not lay their eggs in the water on the plastic or on the plastic itself.

Chapter 31: How to Start Converting to Solar Power

When looking at the usefulness of solar power, you have to consider how much power the appliances that you use will tax the system. For example if you wish to use a solar powered generator to run your hot water, you really don't have to put that much effort into figuring out that a couple of panels will do the trick. But if you want to run a window-type air conditioner or a 24-hour freezer, you would have to consider the possibility of adding more power.

Power-hungry appliances are not conducive to solar power. It can be accomplished, but it is best to rely on the power grid for your larger appliances. Then you can selectively pick the appliances that you want to use with solar power. This can be for only one room. For example in your bedroom you may have a couple of lamps, a television, and a fan. By looking at the amount of energy it takes for running those appliances and the amount of energy that can be produced by a certain number of solar panels, you can make a logical decision to purchase those panels for that use. The rest of the house could run off the grid. As you invest in solar panels, you can slowly convert more rooms to full solar energy.

One good thing about solar panels is that if your house does have an electrical failure due to a storm or high winds, you can go into your solar powered room. You would still be able to watch your television and have lights, without depending on the power grid since it does not supply the power needed for those comforts. Even if your power outage happens at night, the power collected and the solar batteries can run your appliances in the bedroom for several hours or more.

If you want to start converting your house over to solar energy, it is wise to start small unless you have a very big budget. The cost you initially put into a solar powered project will take a long time to turn around and become profitable. Since the Earth's climate is changing, eventually the cost of power could override the cost of the initial set up of solar panels. The long term possibility has to be entered into the equation.

For the first project in converting your home over to solar power, you may want to look at just specific appliances instead of an entire room. Of course, when the power goes out your entertainment is virtually gone. You may want to just use your solar powered energy for your television, cable box, VCR or DVD, or other entertainment equipment. If nothing else, it would be useful in killing the time until the power grid comes back on.

If you are unsure about solar energy and solar power, start small. You can buy a solar panel for your portable devices. Even though this is not saving a great deal of money, it helps the planet when you are not using the power grid to recharge batteries.

You can even buy solar powered panels to power your Game Boy or handheld computer. A computer's battery only lasts two to three hours on the average. If you are away from a power supply, you could be sitting there with a dead box while you could be doing your work, surfing the internet, or dealing with your daily computer tasks.

Chapter 32: Conclusion

The fact is, the 350,000,000 terawatts of power available from the sun is so huge that an exposure to a full sun in only 15 minutes will be enough to generate the world's energy requirement

Compare that with energy that is generated by nuclear and fossil fuel. Presently, the available data for fossil and nuclear fuel is 10,800,000 terawatts which we all know to be non renewable.

To produce electricity, utility companies burn fossil fuels that translate to 1.3 pounds of carbon dioxide to produce 1kw of electrical power. This unwanted CO2 emissions are dumped into the atmosphere. This then translates into each typical home being accountable yearly for 22,000 pounds of CO2 emissions. The harnessing of the sun's rays is clean and safe. It produces no emissions and it is practical and may in the years ahead, prove very economical. In the United States, only 0.1% of power that is generated is solar energy driven. So what are the obstacles?

According to the Wall Street Journal (in an article that was released in its August 2008 issue), there are groups, backed by political groups that are lobbying against the putting up of transmission lines for solar power. The construction of distribution lines for solar energy is also being blocked by environmental activists that restrict the delivering of solar energy to those who want it in their homes.

Another obstacle is that the power grid in the United States which was designed more than 100 years ago is now so congested in many regions. To deliver the solar power to consumers, scientists and engineers will have to come out with another cost efficient plan to transfer huge amounts of energy from one location to another.

Solar panels are considered expensive. Although a home increases its value by folds when solar powered, the costs still could be prohibitive to most that unless the non silicon flexible solar panels that are now being developed are released for market consumption, powering homes through solar energy could still be very limited.

Other forms of rewards to avoid fossil fuel use should still be effectively placed. The 30% tax cut to projected cost previously awarded will be more attractive if other federal credits are included to encourage further investments.

The global warming issue that has been brought to the papers is a recurrent subject of talk shows and remains to be a good news item. Also, the too unstable pump prices, should and for most part, already be a good incentive to use this alternative source of energy.

However, effective solar energy transmission to homes will remain to be very hard unless these obstacles are breached. Assuming that these obstacles are solved today, it will still take some 10 years to convert 20% of American homes into solar energy users. Meanwhile, solar panels on individual homes remain to be the most viable alternative.

The good part to solar energy quest is that technology is advancing very rapidly. Nano technology for solar power is being developed and may be available in five years time. Other breakthroughs in cell designs are also being developed that could, in the next few years, be a cost-effective way of generating energy without having to rely anymore on fossil and nuclear power.